MY FIRST DINOSAUR BOOK

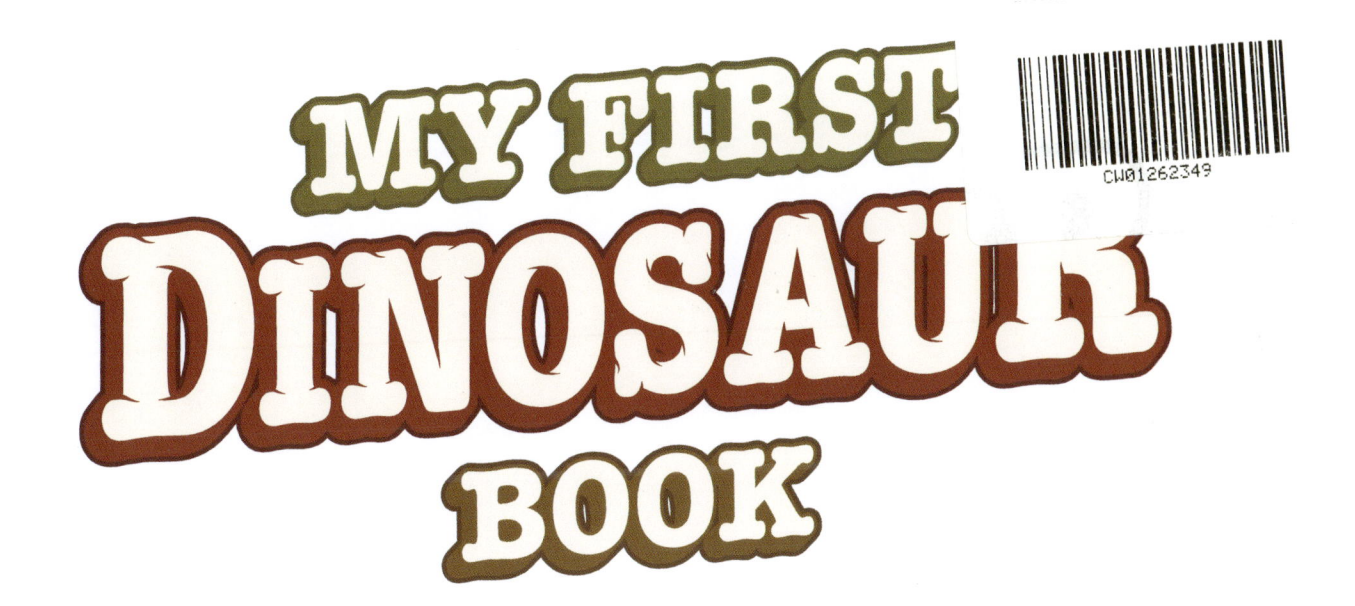

Words & Pictures by
Matt Hazard

Tall Tale

Tall Tale Digital, Inc. — Oakland, CA

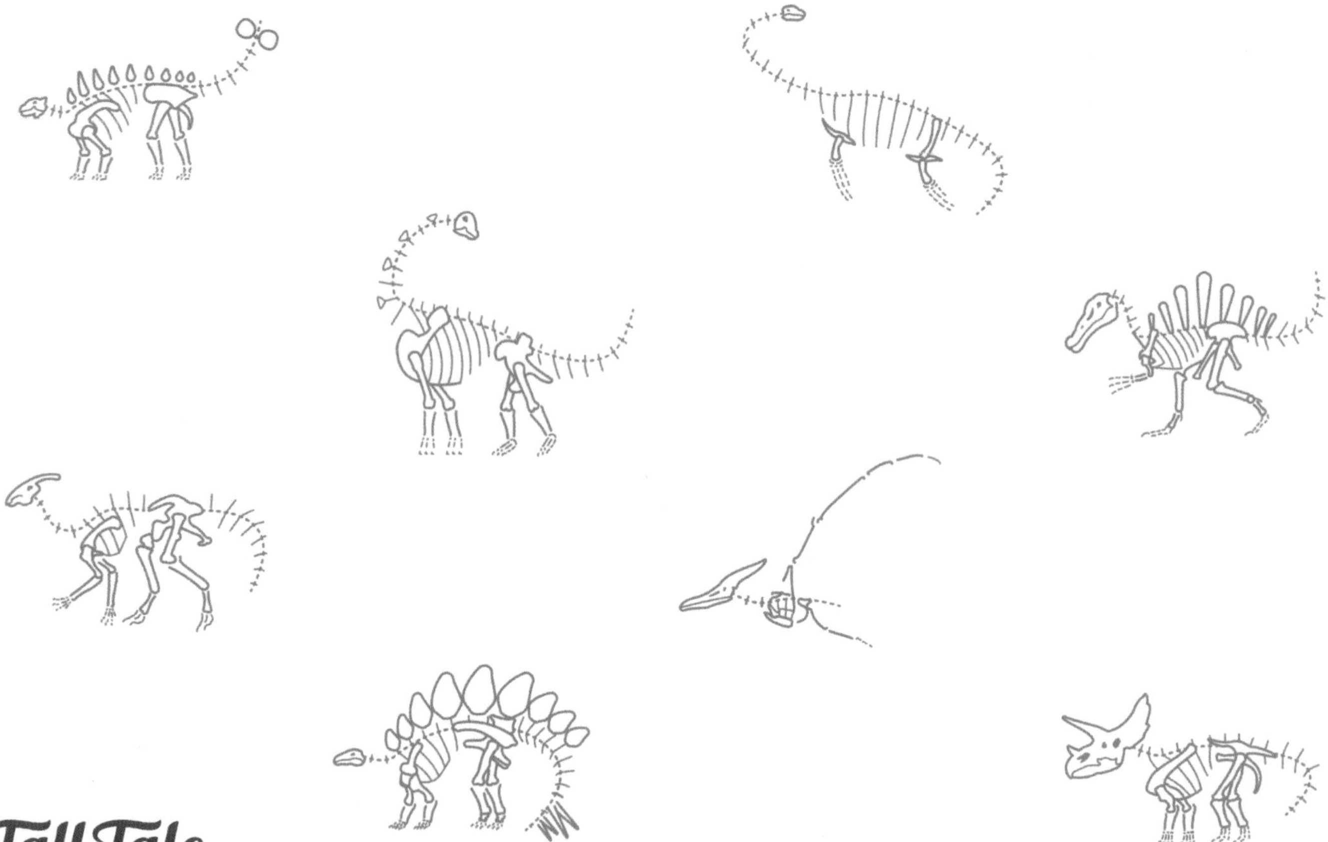

Tall Tale

Tall Tale Digital, Inc. — Oakland, CA

My First Dinosaur Book
Copyright © 2023 Tall Tale Digital, Inc. All rights reserved.

No part of this publication may be reproduced, stored in a retrieval system, or transmitted in any form or by any means, electronic, mechanical, photocopying, recording (including but not limited to storytime videos), or otherwise, without written permission from the publisher, except in the case of brief quotations embodied in critical articles and reviews. For information regarding permission, write to Matt Hazard at Tall Tale Digital, Inc. by visiting www.talltaledigital.com

ISBN:
978-1-955460-02-6

LCCN:
2023918050

This book belongs to

....

DINOSAURS

were reptiles that lived between 251.9 to 66.0 million years ago in the **Mesozoic Era**. The three parts of that era were the **Triassic**, **Jurassic**, and **Cretaceous Periods**.

Today, we can see dinosaur bones called **fossils** in a museum. The two main groups of dinosaurs were plant-eaters called **herbivores**

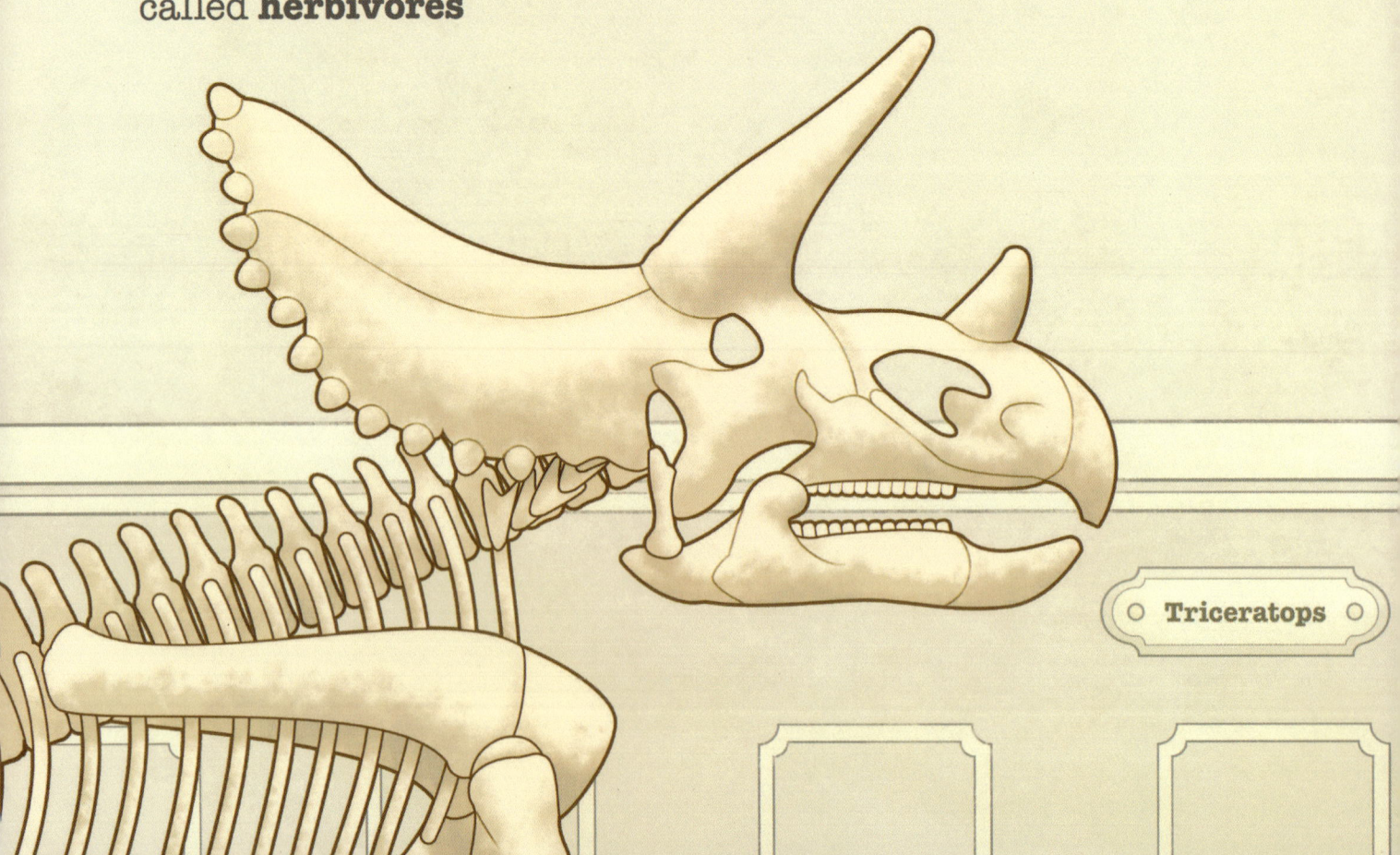

Triceratops

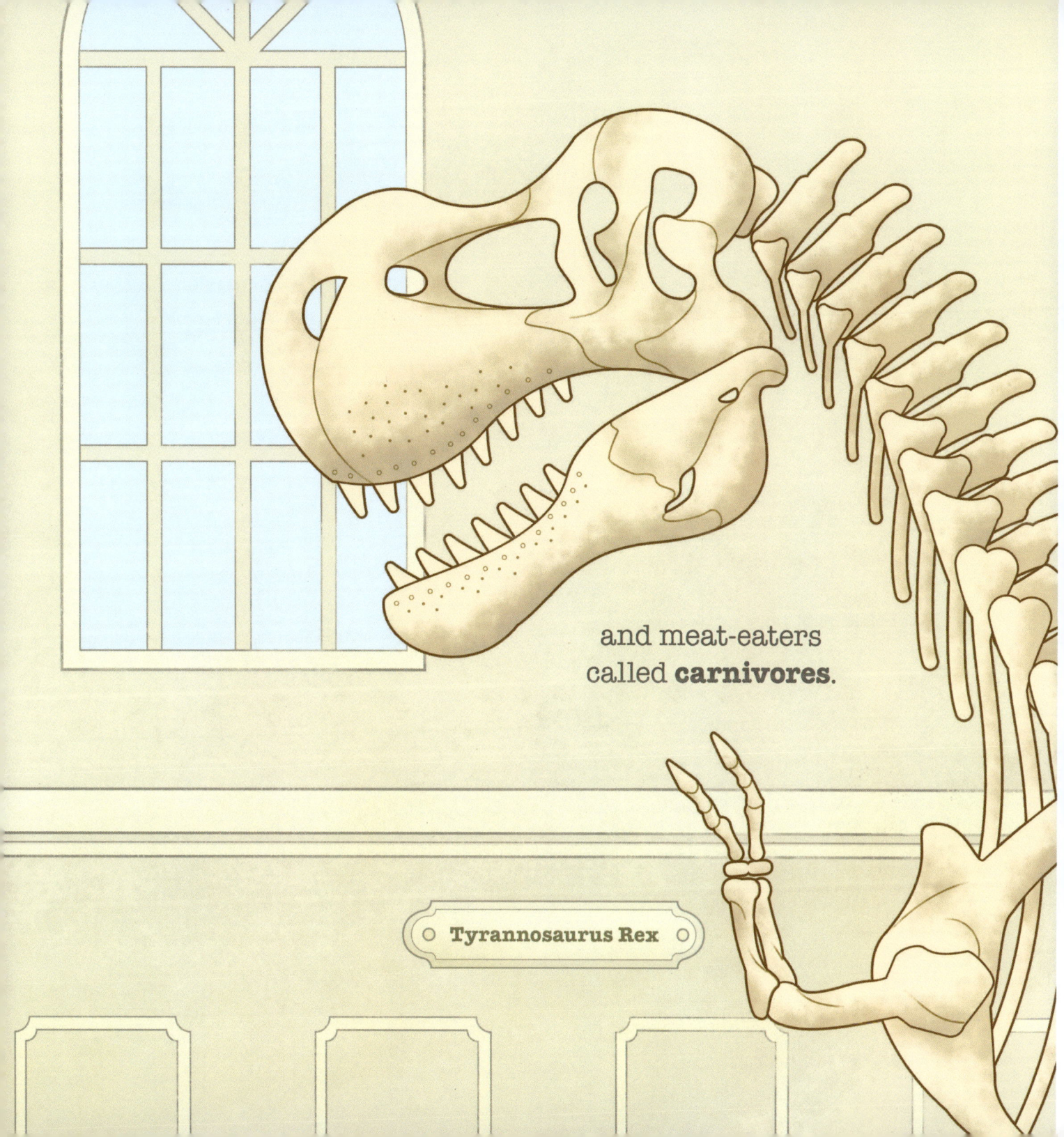

and meat-eaters called **carnivores**.

Tyrannosaurus Rex

Hi, I'm Paulo, a third-grader who loves dinosaurs. I'm a junior **paleontologist** at the natural history museum.

Paleontologists study prehistoric plants and animals. They discover dinosaurs by digging up **fossils** from the ground.

Let's tour the museum together to learn more about some of my dinosaur pals!

Anna, the
ANKYLOSAURUS
an-KIE-loh-SORE-us
means "fused lizard"

Anna was an **herbivore** that used a club-like tail and spiky shell for protection from carnivores.

Anna lived in the **Cretaceous Period**

Ankylosaurus was up to 30 feet long from nose to tail and weighed up to 5 tons.

Brady, the
BRACHIOSAURUS

BRAK-ee-oh-SORE-us
means "arm lizard"

Brady was a giant **herbivore** that used a very long neck to reach leaves from tall trees.

Brady lived in the **Jurassic Period**

Brachiosaurus was up to 49 feet tall, 85 feet long from nose to tail, and weighed up to 55 tons.

Eliza, the
ELASMOSAURUS

eh-LAZZ-mo-SORE-us
means "thin-plated lizard"

Eliza was a **carnivore** that lived in the ocean, feeding on smaller dinosaurs and fish.

Eliza lived in the
Cretaceous Period

Elasmosaurus was up to 46 feet long from nose to tail and weighed up to 10 tons.

Parker, the
PARASAUROLOPHUS

pa-ra-saw-ROL-off-us
means "near-crested lizard"

Parker was an **herbivore** called a duck-bill with a long bony crest that might have sounded like a foghorn.

Parker lived in the **Cretaceous Period**

Parasaurolophus was up to 36 feet long from nose to tail, with a 5-foot crested head, and weighed up to 4 tons.

Teresa, the
PTERANODON

TERR-ran-OH-don
means "winged-toothless lizard"

Teresa was a **carnivore** with thin, hollow bones and could fly like a bird.

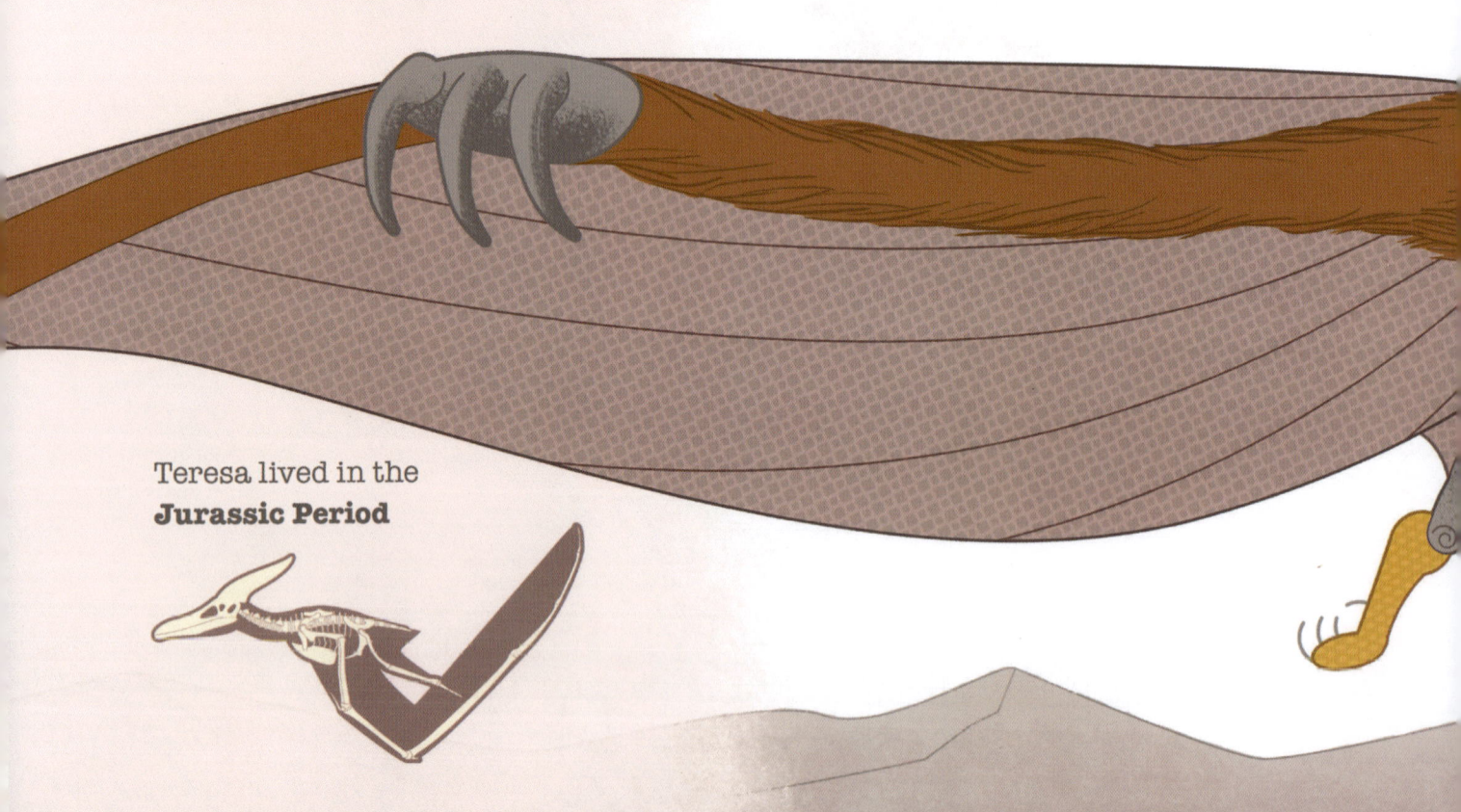

Teresa lived in the
Jurassic Period

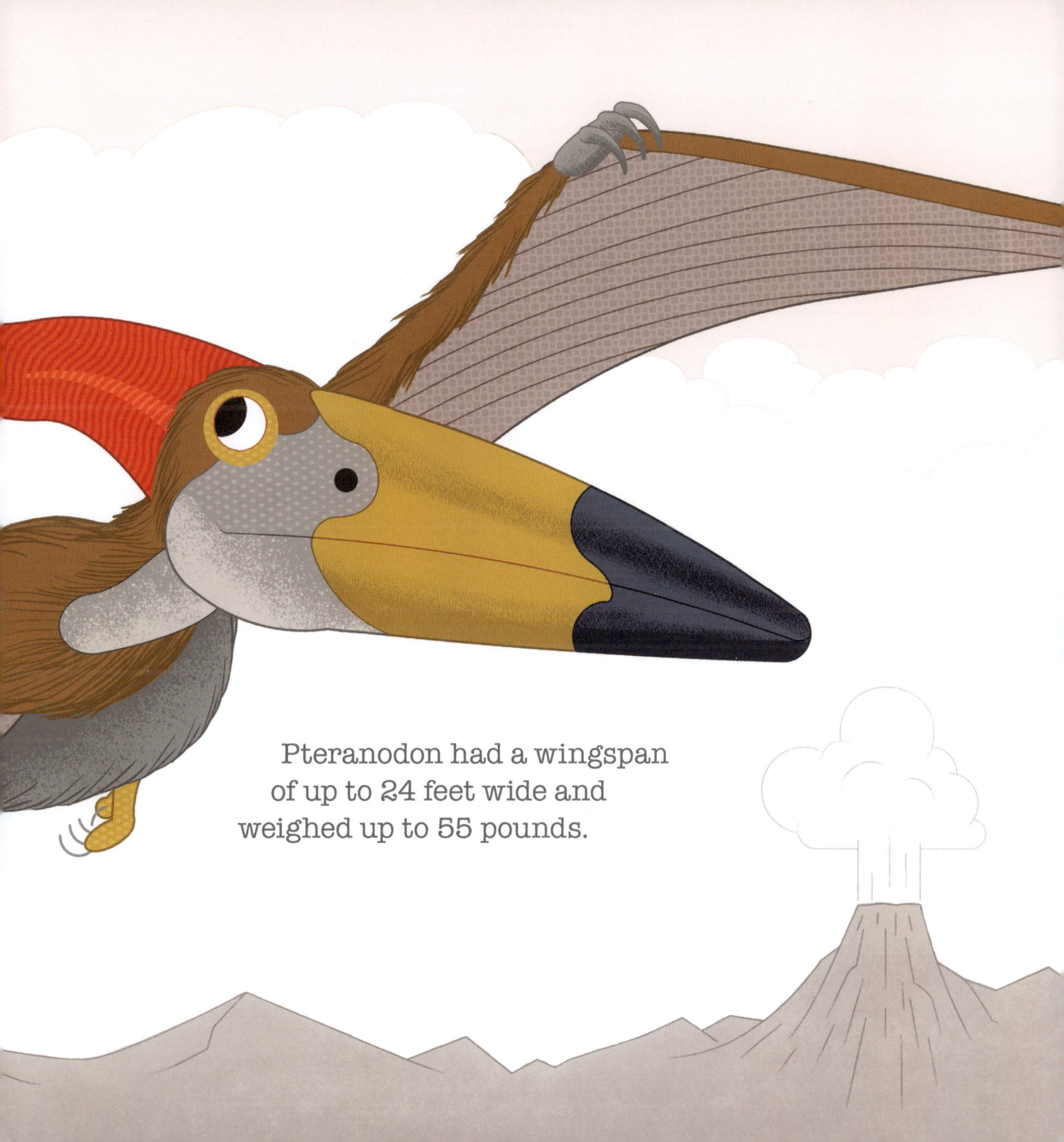

Pteranodon had a wingspan of up to 24 feet wide and weighed up to 55 pounds.

Spencer, the
SPINOSAURUS

SPINE-oh-SORE-us
means "spine lizard"

Spencer swam like a crocodile and might have been the biggest **carnivore**.

Spencer lived in the **Cretaceous Period**

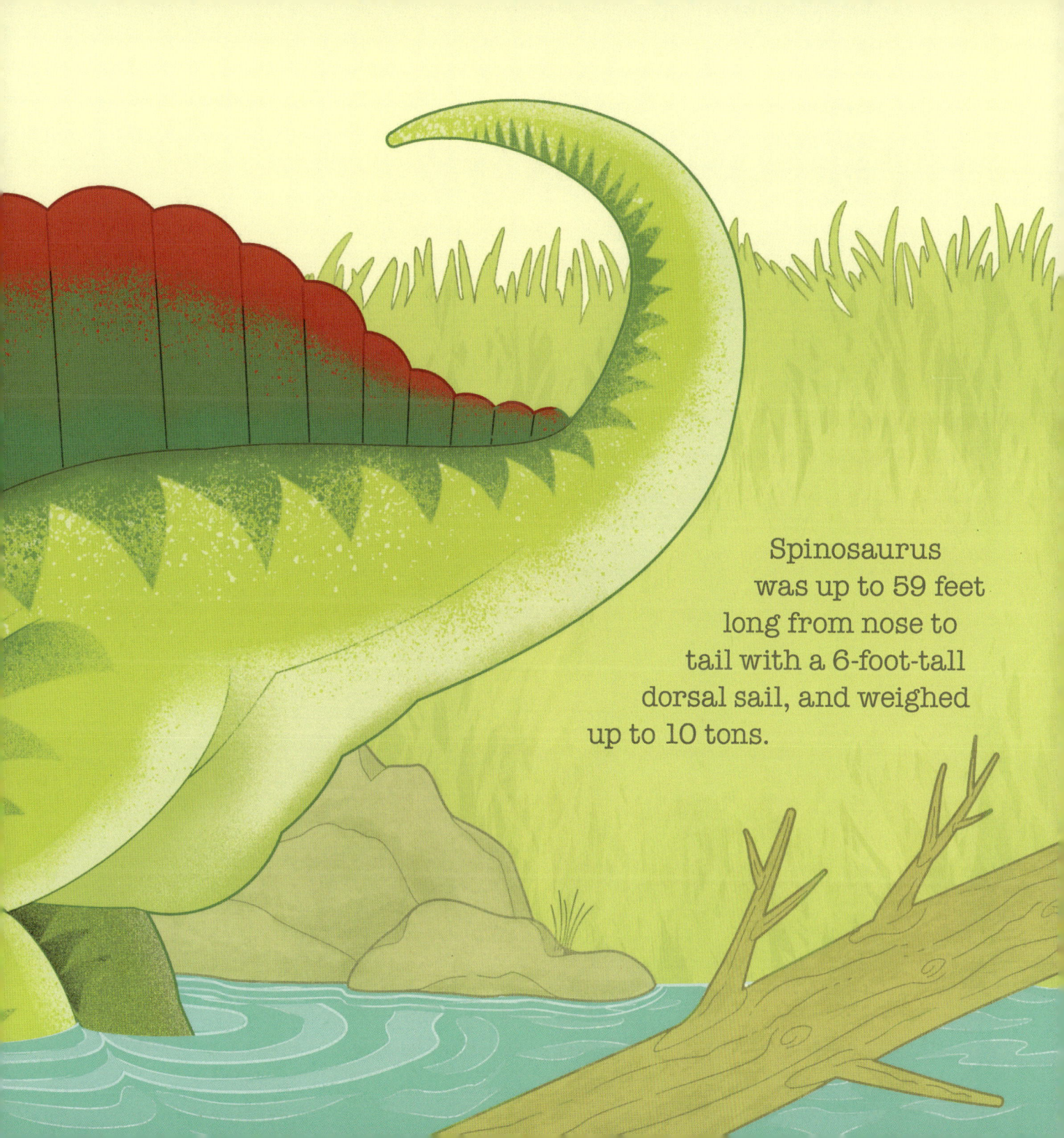

Spinosaurus was up to 59 feet long from nose to tail with a 6-foot-tall dorsal sail, and weighed up to 10 tons.

Stella, the
STEGOSAURUS

STEG-oh-SORE-us
means "roof lizard"

Stella was an **herbivore** with 3-foot-tall backplates that might have been for keeping her body temperature just right.

Stella lived in the
Jurassic Period

Stegosaurus was up to 30 feet long from nose to tail and weighed up to 7 tons.

Tristan, the
TRICERATOPS

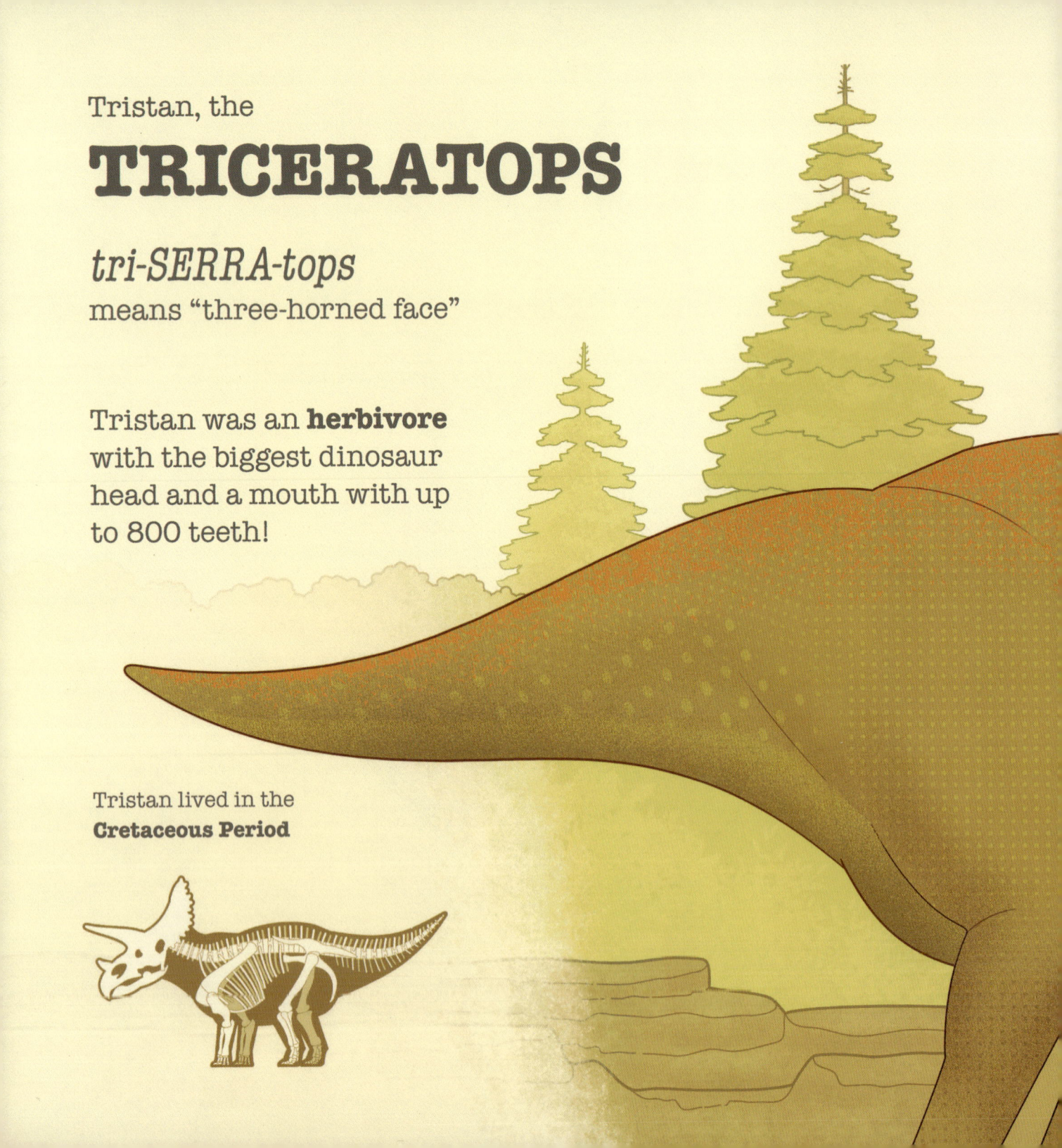

tri-SERRA-tops
means "three-horned face"

Tristan was an **herbivore** with the biggest dinosaur head and a mouth with up to 800 teeth!

Tristan lived in the
Cretaceous Period

Triceratops was up to 30 feet long from nose to tail with an 8-foot skull, and weighed up to 8 tons.

Tyra, the
TYRANNOSAURUS REX

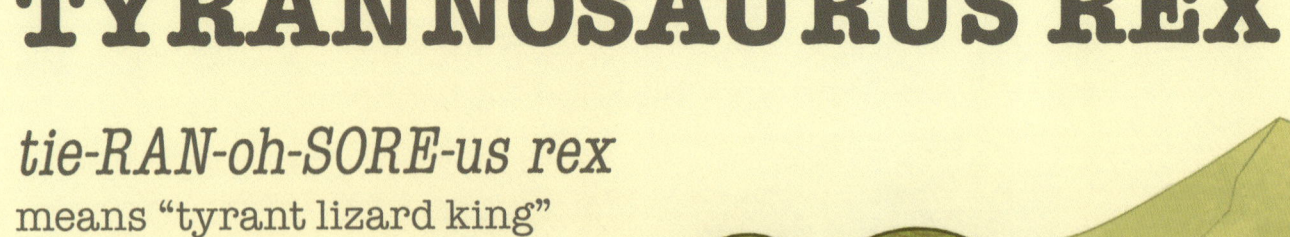

tie-RAN-oh-SORE-us rex
means "tyrant lizard king"

Tyra was a **carnivore** that had teeth the size of bananas and could eat 200 pounds in one bite!

Tyra lived in the **Cretaceous Period**

Tyrannosaurus Rex
was up to 40 feet long
from nose to tail,
up to 20 feet tall, and
weighed up to 8 tons.

Vern, the
VELOCIRAPTOR

vel-OSS-ee-rap-tor
means "quick thief"

Vern was a turkey-sized **carnivore** that may have been covered in feathers and hunted in a pack like a wolf.

Vern lived in the **Cretaceous Period**

Velociraptor was up to 6.5 feet long from nose to tail and weighed up to 100 pounds.

What did you learn?

- Which scientists study prehistoric plants and animals?
- What are dinosaur bones called?
- What kind of food did herbivores eat?
- What kind of food did carnivores eat?
- What were the three parts of the Mesozoic Era?

Size of a grownup

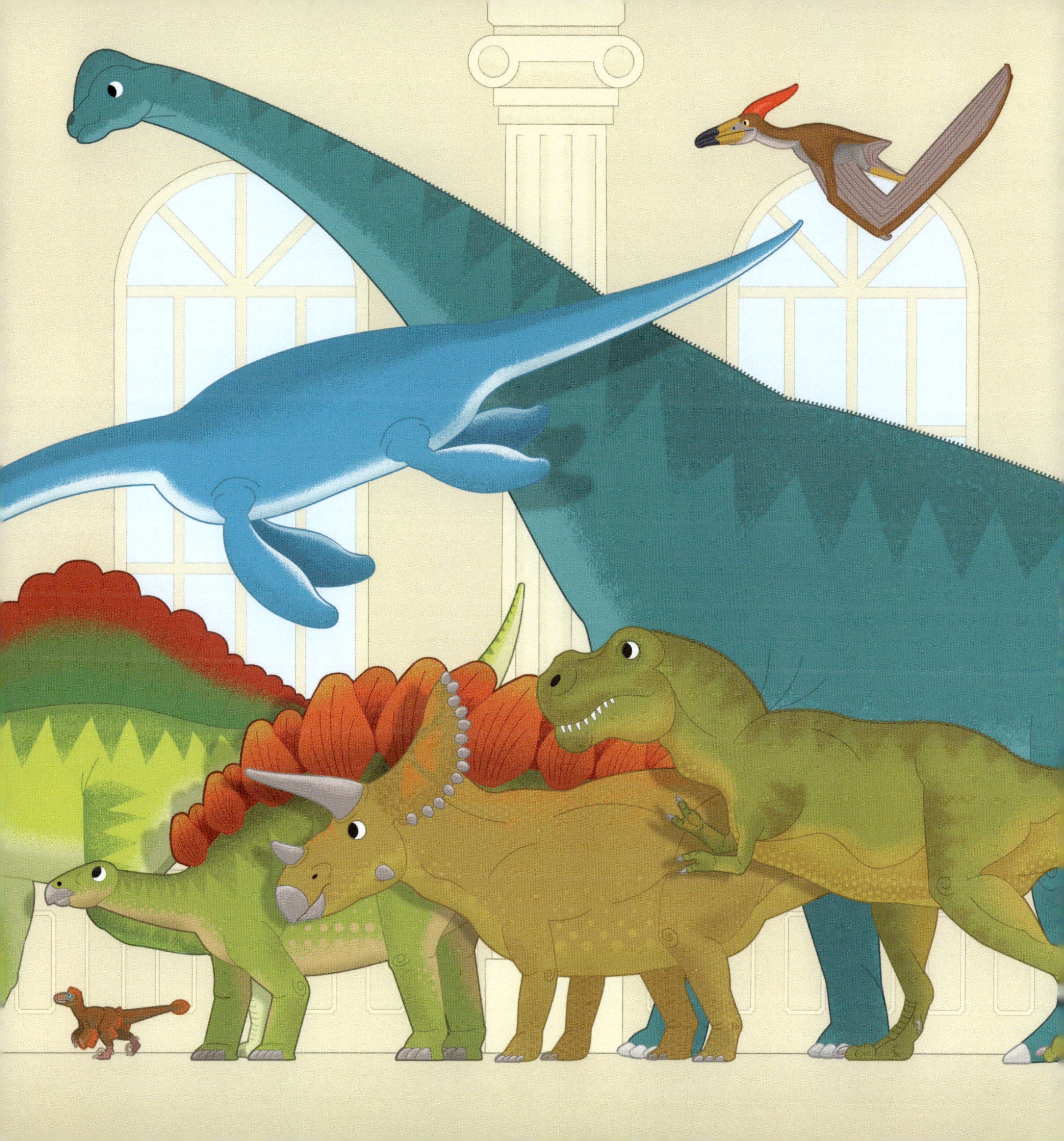

Parents, to get free printable activities, watch animated videos, and discover other books and apps, visit talltaledigital.com using this QR code:

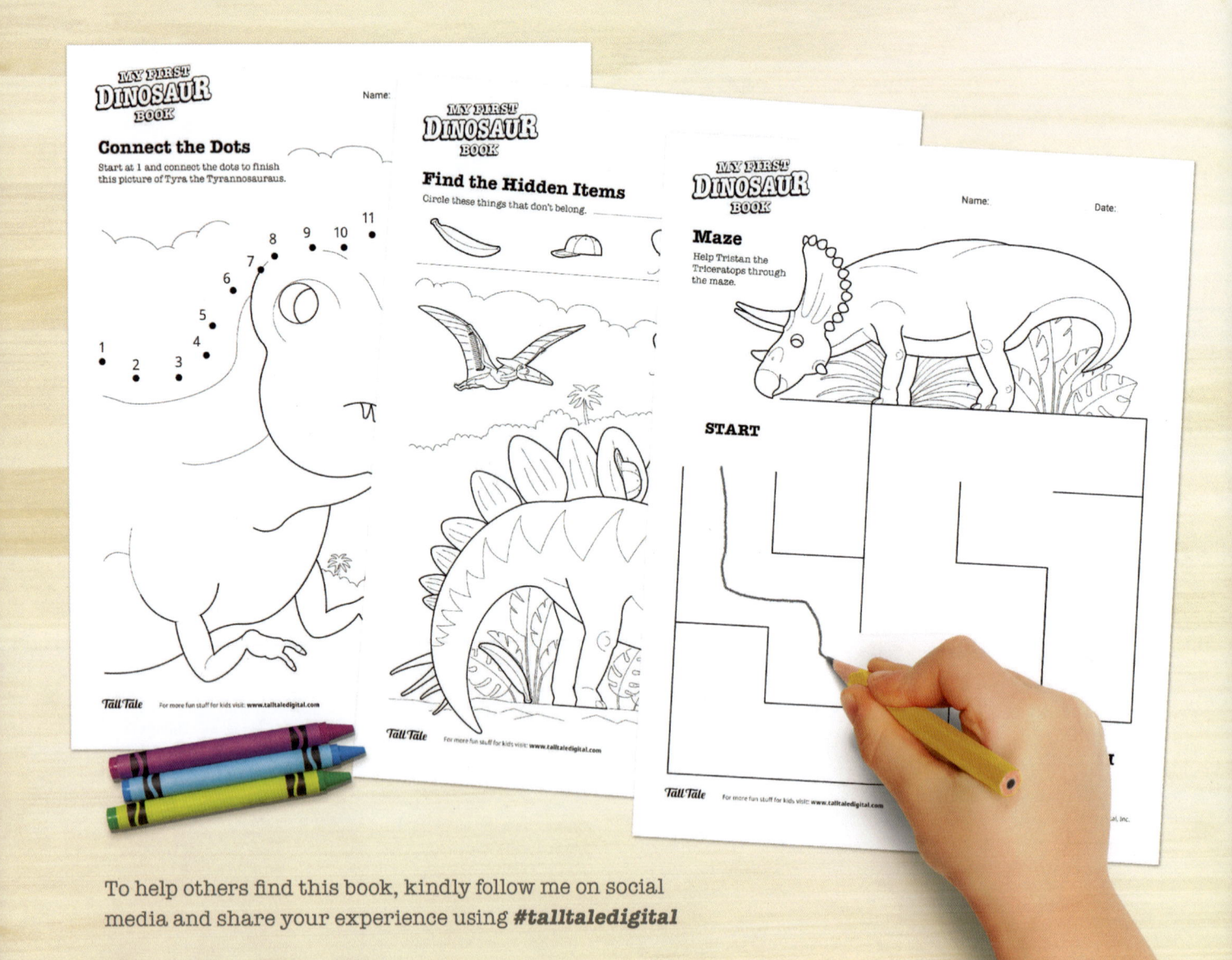

To help others find this book, kindly follow me on social media and share your experience using #talltaledigital

About Matt

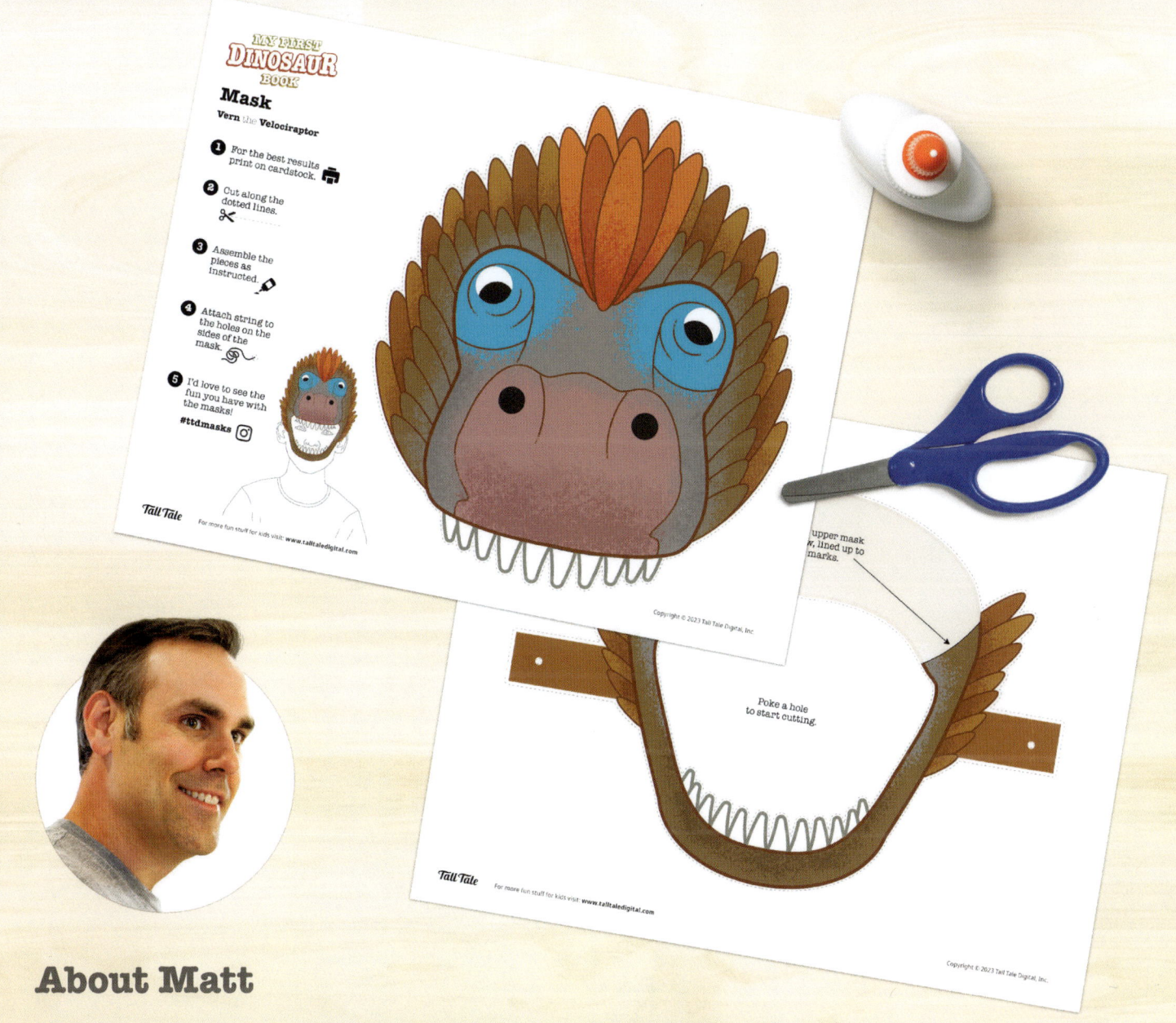

Matt Hazard has been designing user experiences, making animated videos, and illustrating for over twenty years. He makes activities for kids that reinforce educational concepts, creative expression, and social norms such as sharing, empathy, and cooperation. He writes and illustrates material for his books, apps, and other activities in his studio in Oakland, CA, where he lives with his wife, daughter, and golden retriever.

Printed in Great Britain
by Amazon